ANIMALS OF THE NIGHT

Design
David West
Children's Book Design
Illustrations
Stella Robinson
Picture Research
Cecilia Weston-Baker
Editor
Scott Steedman
Consultant
Miles Barton

© Aladdin Books Ltd 1989

Designed and produced by
Aladdin Books Ltd
70 Old Compton Street
London W1

*First published in the
United States in 1989 by*
Gloucester Press
387 Park Avenue South
New York, NY 10016

Printed in Belgium

Library of Congress Cataloging in Publication Data

Bender, Lionel
    Animals of the night / by Lionel Bender.
        p. cm. – (First sight)
    Includes index.
    Summary:Explains why some animals are more
active at night and the physical characteristics, habits,
and natural environment of such nocturnal animals as
bats, frogs, toads, marsupials, and others.
    ISBN 0-531-17161-2
    1. Nocturnal animals – Juvenile literature. [1.
Nocturnal animals.] I. Title. II. Series.
OL755.5.B46 1989
591.5'3–dc20   89-31559
                                                             CIP
                                                             AC

This book tells you about animals that are
active at night – where they live, what they eat
and how they survive. Find out some surprising
facts about them in the boxes on each page.
The identification chart at the back of the book
will help you when you see night creatures in
zoos or in the wild.

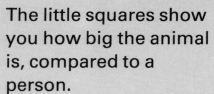
The little squares show
you how big the animal
is, compared to a
person.

A red square means that
the animal is endangered
in part or all of its range.
See the survival file.

The picture opposite shows a Bushbaby, an African night animal

FIRST SIGHT

ANIMALS OF THE NIGHT

Lionel Bender

GLOUCESTER PRESS

New York · London · Toronto · Sydney

Introduction

As night falls many animals, including people, go to sleep or to rest in their homes. But other animals are just beginning their survival routines. These animals of the night are said to be "nocturnal." By being active during the hours of darkness, they make use of food sources and habitats that are used by other animals when the sun is out. They are also safe from predators that hunt during the day.

Most nocturnal creatures are highly adapted to life with little or no light. They can find their way, locate food and detect dangers in almost total darkness.

◁ **Fennec Foxes do most of their exploring at night**

Special senses

Nocturnal animals have at least one sense – sight, hearing, smell, taste or touch – which is highly developed. These powerful senses help them to survive in the dark. Tigers, for example, have excellent eyesight and hearing and a keen sense of smell. Their eyes face forward, allowing them to judge distances accurately when they hunt at night.

Rattlesnakes and pythons can locate their prey in total darkness. They do this by using the heat-sensitive pits on their faces. The pits help them detect small animals, such as rats and birds, which are warmer than their surroundings. These heat sensors can tell temperature differences of one-hundredth of a degree Fahrenheit.

Male moth with large, sensitive antennae

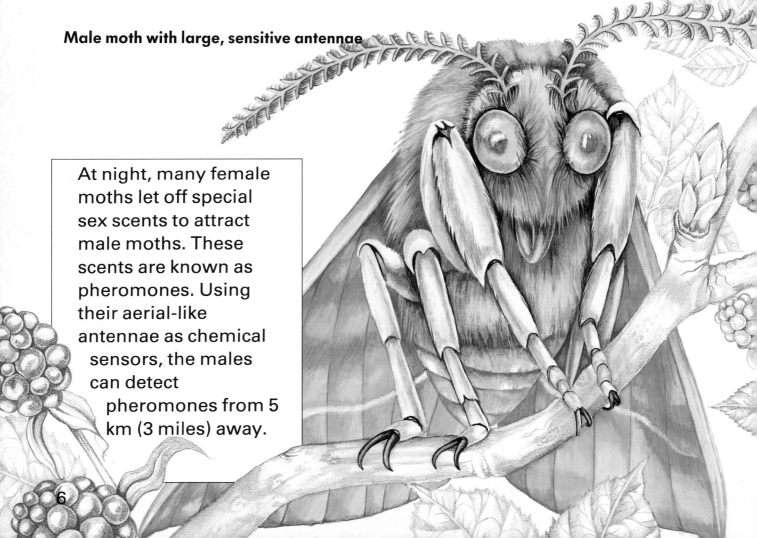

At night, many female moths let off special sex scents to attract male moths. These scents are known as pheromones. Using their aerial-like antennae as chemical sensors, the males can detect pheromones from 5 km (3 miles) away.

Nightjars are night-time predators. They use a system called "echolocation" to find their prey. The bird emits sounds as it is flying along. These sounds strike an insect and bounce back. The nightjar listens for the echoes, which allow it to determine the insect's position accurately.

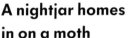

A nightjar homes in on a moth

Aerial hunters

Owls live in woods and forests. They rest in trees during the day. But as darkness falls they take to the air and begin to hunt. Owls feed on mice, voles and small birds. Sitting on a perch, a Barn Owl will listen for the slightest squeak or rustle from the ground below. To do this it moves its head from side to side, up and down, and almost right around. When it detects an animal the owl flies towards it, using its sharp vision to home in on its target. Owls' eyes are up to a hundred times more sensitive to light than ours, and they have the widest field of vision of any bird.

Nightjars and nighthawks hunt insects at night. They open their enormous mouths as they fly and gobble up mosquitoes and moths in their path. Some will even snap up other birds if they come across them.

A Barn Owl swooping in for the kill ▷

Scavengers and predators

Many predators – animals that kill and eat other animals – hunt at night. Civets, genets and mongooses, for example, are nocturnal predators. The animals they hunt, their prey, include birds, mice and insects. Leopards and tigers stalk their antelope prey in the darkness. Hyenas are efficient night-time killers, but they are also scavengers – they feed on the leftovers of other animals' meals. A pack of hyenas will even drive a lion away from a carcass.

Wild boars mainly eat plant food, and skunks feed on insects, mice, bird and reptile eggs, and fruit. But these nocturnal animals also feed on the flesh of any dead animals that they find. If a skunk is threatened, it stands on its front paws and squirts a jet of foul-smelling liquid at the intruder's eyes. The liquid causes burning and inflammation.

Spotted Hyenas try to bring down a young zebra

A tiger moves its kill to a safe place ▷

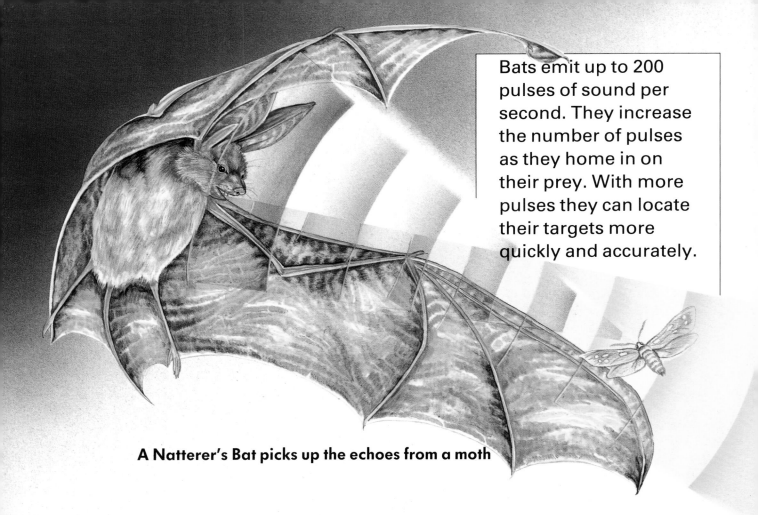

Bats emit up to 200 pulses of sound per second. They increase the number of pulses as they home in on their prey. With more pulses they can locate their targets more quickly and accurately.

A Natterer's Bat picks up the echoes from a moth

Bats

Bats spend the daytime sleeping in caves, trees or old buildings. They sleep hanging upside-down, sometimes in huge groups of a million or more. They start to search for food soon after the sun has set. Most bats feed on insects, which they capture while in flight. They have small eyes and poor vision – hence the saying "as blind as a bat."

Like nightjars, bats use echolocation to pin-point insects in the air. Horseshoe Bats, for example, have flaps of skin on the nose. They use these to direct sounds which they fire out of their nostrils. They send out many pulses of sound each second and listen for any echoes. Bats have very good hearing. Their large, curved ears collect sounds from over a wide area. Bats that live in caves also use echolocation to find their way around in dark, cramped spaces.

Brown Bats use echolocation when flying ▷

Avoiding the heat

Snakes, lizards and all other reptiles are "cold-blooded." Their body temperature depends on their surroundings. Many reptiles that live in deserts are nocturnal. They avoid the heat and bright light of the day by burrowing or sleeping in the shade, and only come out when darkness falls. Pit vipers like the rattlesnake, for example, rely on heat from sun-warmed rocks and soil. They hunt nocturnal mice, frogs, lizards, and sleeping birds at night.

Many desert mammals burrow in the ground to escape the fierce heat. They include jerboas, gerbils and some rats and mice. Desert jerboas feed on insects and plants. They have large eyes, giving them good night vision, and their big ears can detect the faintest of sounds. They feed on insects and plants and often carry food back to their burrows to eat in safety or store until later.

These desert animals all escape the midday sun by digging burrows in the ground

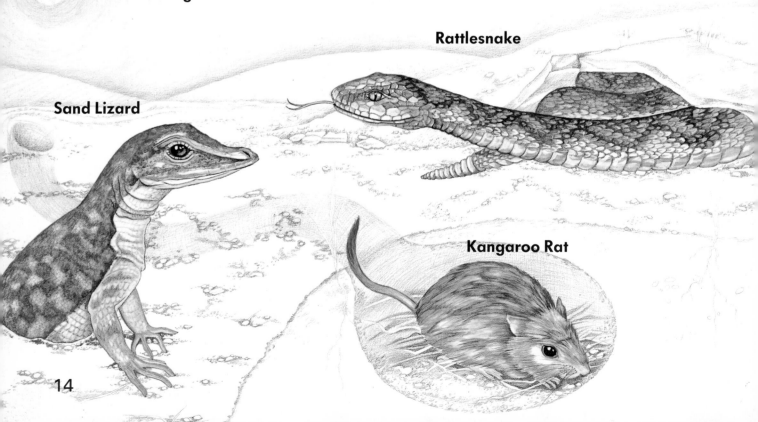

Rattlesnake

Sand Lizard

Kangaroo Rat

14

Spiders and scorpions

As night falls a Trapdoor Spider will gently raise the silken door at the entrance to its burrow. Any insect that comes too near is seized by a lightning swoop of the spider's front legs. The prey is then dragged into the burrow and eaten. Bird-eating Spiders are also night-time hunters. They are covered in hairs that are highly sensitive to vibrations. Occasionally they catch birds, but usually they prey on frogs, insects and other spiders. They can give a person a very painful bite.

Scorpions hunt spiders, centipedes and insects at night. They wait in their lairs for their victims to pass by. Then they strike. They catch their prey with their pincer-like claws, and paralyze or kill it using the sting at the end of their tail. During the day scorpions live in hollows beneath rocks or in burrows up to 1 m (3 ft) deep.

Trapdoor made of silk

Ant

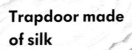
Trapdoor Spider

Spider ready to pounce

This grasshopper strayed too close to a scorpion's den ▷

Tarsiers are only the size of squirrels. But they can leap 2m (6.5 ft) from branch to branch in utter darkness. They almost never fall.

In bright light, a tarsier's pupils will close up, as here

Primates

Tree-shrews, lorises, pottos, and tarsiers closely resemble our distant ancestors. Together with monkeys, apes, and humans, they form a group of mammals known as primates. And like the early primates, most of them are small tree-living animals that are active at night. They spend the day sleeping, either inside a hollow tree or clinging to a branch with their hands and feet. They eat fruit, leaves, insects and even frogs.

Tarsiers and lorises have large, forward-facing eyes. These help the animals to judge distances accurately as they move from tree to tree in the starlit forests. Their hearing is also good. Like bats, they have large ears. They move these constantly to locate the sources of sounds. Slow Lorises spend their entire lives in trees. Females make an eerie whistle at night. This is believed to be a mating call used to attract male Slow Lorises.

Slow Lorises live in the rain forests of southern Asia ▷

Urban visitors

Many animals of the night make their homes close to human settlements. Parks and gardens provide refuge for snails, slugs and earthworms. These creatures live in damp, shady places among the leaf litter or in the soil. They emerge at night or on wet days and go searching for plant food and decaying material. Because they move slowly and have few defenses, many are eaten by badgers and moles. These nocturnal predators use their keen senses of smell and hearing to find their prey.

Raccoons and Red Foxes also eat earthworms. But they prefer to prey on mice, frogs, fish, and birds. In parts of North America and Europe they are thought of as pests. They raid garbage cans and have been known to kill cats, chickens, and even lambs.

A North American (above) and two European night-time garden visitors

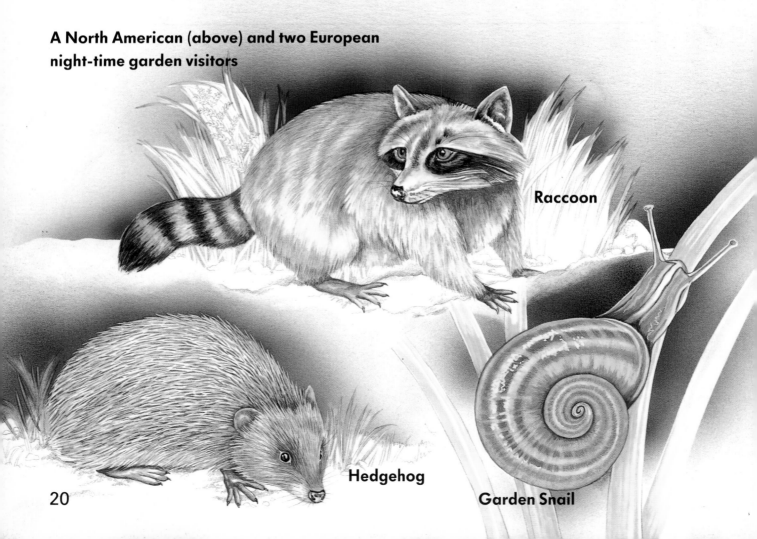

Raccoon

Hedgehog

Garden Snail

In a split-second, a frog can shoot out its long and sticky tongue, catch an insect in flight, and pull the prey into its mouth.

Frogs and toads

Spring is the mating season for frogs and toads in the temperate parts of the world. The night air is filled with their croaks, chirps and grunts. All frogs and toads are more active at night. This is because they are safer from attack by reptiles and birds, most of which hunt in the day. There is also little danger of their moist skin drying out at night, and there are plenty of insects to eat.

Frogs and toads mainly rely on vision to catch their food. Even in very dim light, the slightest movement of a small animal will draw their attention. Their eyes bulge out from the head, giving them a clear, all-around view of food or approaching danger.

Male frogs croak to tell females where to find them ▷

Insect eaters

"Aardvark" means "earth pig." It is an appropriate name, for this noctural grassland animal is about the size of a large pig. It uses its long mobile snout to root around in the earth for food. The Aardvark mostly feeds on ants and termites. Its sense of hearing is very good: it can detect a column of ants on the move from the rustling noise made by the insects.

Aardvarks are powerful animals. They rip open termite nests with their claws and lap up the insects with their long sticky tongues. Pangolins feed in the same way, but they seem to use smell rather than hearing to locate their food. When threatened, they curl up into a tight ball. Their sharp-edged scales protect them from attackers. Tree Pangolins spend most of their time high in the forest canopy, feeding from ant and termite nests.

Tree Pangolins are almost blind, but have strong senses of smell and hearing

An Aardvark sleeps after its night-time feast of termites ▷

Pouched animals

Animals like kangaroos and the Koala are mammals, as we are. But unlike us, they give birth to tiny young that develop in a pouch outside their mother's belly. Animals that do this are known as marsupials. Most marsupials are plant-eaters and many are nocturnal.

Possums move through forests under cover of darkness, traveling between feeding and resting sites. They have good senses of hearing and smell. Possums mark their areas with their urine and dung, and rub one another with scents from skin glands. Marsupial meat-eaters, such as the Tasmanian Devil and Australian Native Cat, have good sight, hearing, and smell. They are excellent hunters. They prey on birds, lizards, insects, rats, and rabbits. Some are also scavengers.

A Tasmanian Devil shows its teeth

This Virginian Opossum is carrying her young on her back ▷

Survival file

Animals of the night are many and varied. Some, like the primates and several opossums, are threatened with extinction. This may be through destruction of their forest homes for timber and to create new farmland. Or it may be a result of hunting by local people for their skins and meat. Other nocturnal animals, such as the Common Raccoon, are thought of as pests. They often venture into towns and cause chaos by raiding garbage dumps and attacking pets. But as with all nocturnal animals, because they are most active when it is dark, we know little about their behavior. This makes it more difficult to manage and protect them.

Kangaroos slaughtered for their meat

Nocturnal creatures such as tigers and leopards are now the subject of international conservation agreements. For centuries, these animals were killed for "sport" or to provide fur coats or rugs. Fifty years ago there were still about 30,000 Indian Tigers remaining. Today there are less than 3,000, though numbers have been increasing since the start of the Save the Tiger Campaign in the 1970s, and the creation of reserves.

Owls need barns like this one to roost in

A Leopard skin for sale

Sometimes nocturnal (and daytime) creatures are threatened because of harmful diseases. The European Badger, for example, may carry the disease tuberculosis. Farmers then kill sick Badgers to stop the infection from spreading to cattle. In North America the Common Raccoon may carry another disease, rabies, which can be fatal to pets – and to people. An infected animal must be killed if the spread of the disease is to be prevented.

Kangaroos and wallabies, many of which are nocturnal, are often killed deliberately to keep their numbers under control. These animals are regarded as pests because they eat grass needed for sheep and goats. Many species of bats and owls have suffered from our increasing use of pesticides, and from the removal of farm buildings in which they roost and nest.

A Red Fox on a raid

Identification chart

This chart shows you a variety of large and small animals of the night. There are mammals, birds, reptiles, amphibians and invertebrates from different regions of the world. Notice how the larger animals all have well-developed sense organs – large eyes, a long snout with touch-sensitive whiskers, or large ears. These help the animals to find their way, and their food, in the dark. You can see most of these animals in zoos, and a few of them in your yard.

- Europe
- Australia
- Africa N.
- N. and S. America
- Madagascar
- Asia

European Hedgehog

Aardvark

Kangaroo

Rattlesnake

European Badger

Gecko

2

Make a night-time animal mural

1. Using the drawings above, copy the outlines of the animals onto sheets of graph paper.
2. Tape a sheet of aluminum foil to the back of each outline.
3. Cut out the animal shapes from the foil.
4. Stick the shapes onto a large sheet of black paper.
5. Hold up the large sheet in front of a flashlight in a dark room and make the animals glow.

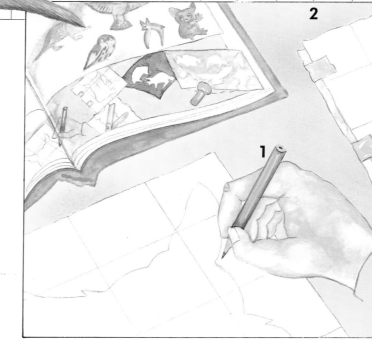

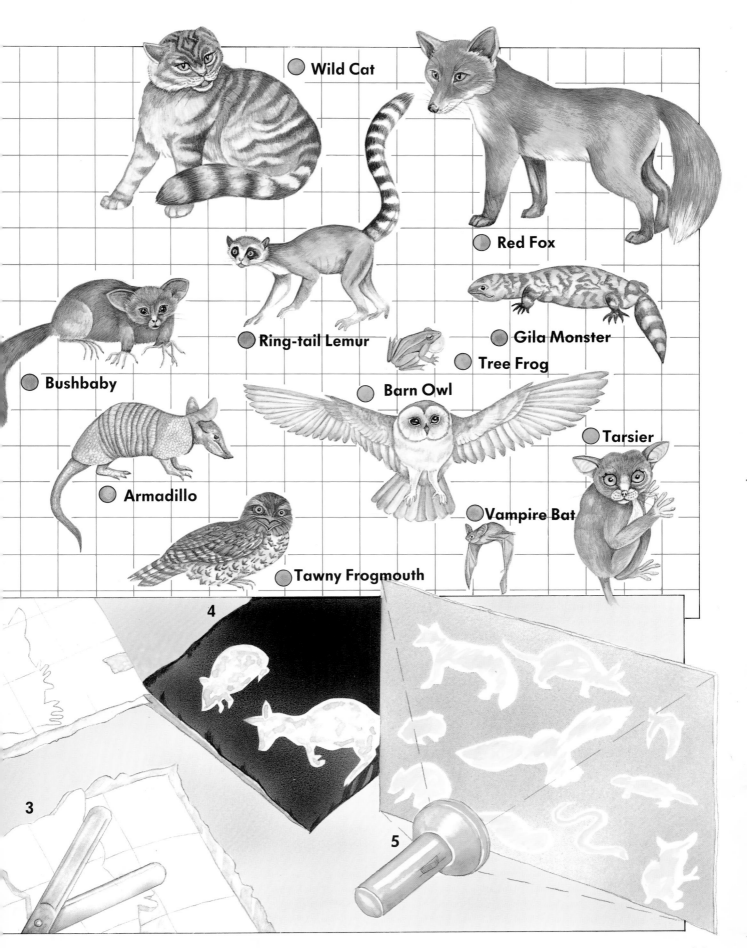

Wild Cat

Red Fox

Ring-tail Lemur

Gila Monster

Tree Frog

Bushbaby

Barn Owl

Armadillo

Tarsier

Vampire Bat

Tawny Frogmouth

4

3

5

Index

Photographic Credits:
Cover, title page and pages 4, 7, 13, 15, 17, 19, 21, 23, 24, 25, 26, 27, 28 and 29 all: Bruce Coleman; pages 9 and 11: Planet Earth.